Ron Mueller

Praise for
Stress Free™ Maintenance Solutions

Ron Mueller

Stress Free^{TM} Maintenance Solutions
for
Manufacturing and Production Systems
By: Ron Mueller

Around the World Publishing LLC
4914 Cooper Road Suite 144
Cincinnati, Ohio 45242-9998

ISBN: 978-1-68223-254-5
ISBN: 1-68223-254-9

Distribution by: Ingram
Cover Picture by: Andrey Popov, Dreamstime.com
Cover Design by: Ron Mueller

Technical Editor:

Gordon Miller P. E.

DEDICATION

To all the dedicated people that strive to optimally maintain the equipment that transforms raw materials into a finished product.

Table of Content

Overview

Stress Free™ Maintenance Solutions is a system of maintaining and improving the production, safety and quality systems by stabilizing equipment and processes and by enhancing the skills of all the people involved in maintaining the equipment and systems.

Equipment maintenance is any process used to keep a business's equipment in reliable working order. It includes routine upkeep as well as corrective repair work and periodic rebuild.

Maintenance includes mechanical assets, tools, and computer systems.

The measurable objectives are optimum.

- OEE
- Throughput

And minimum.

- Cost
- Effort

Planned Maintenance System

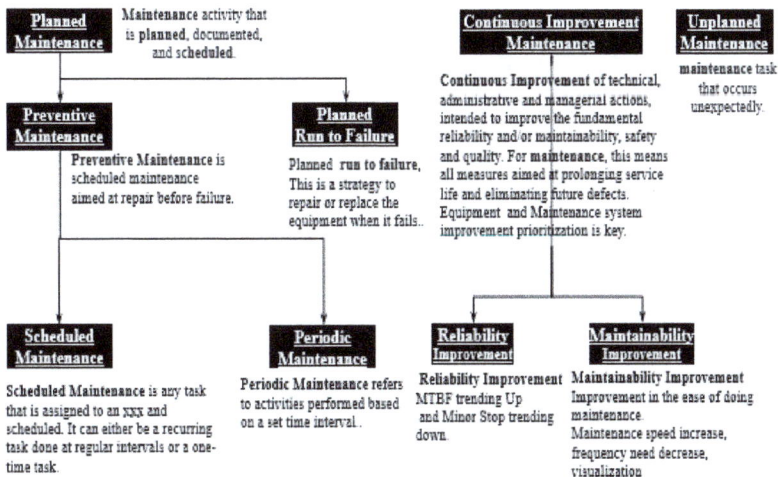

| Planned Maintenance | Maintenance activity that is planned, documented, and scheduled. | | Continuous Improvement Maintenance | Unplanned Maintenance |

Planned Maintenance — Maintenance activity that is planned, documented, and scheduled.

Preventive Maintenance — Preventive Maintenance is scheduled maintenance aimed at repair before failure.

Planned Run to Failure — Planned run to failure, This is a strategy to repair or replace the equipment when it fails..

Continuous Improvement Maintenance — Continuous Improvement of technical, administrative and managerial actions, intended to improve the fundamental reliability and/or maintainability, safety and quality. For maintenance, this means all measures aimed at prolonging service life and eliminating future defects. Equipment and Maintenance system improvement prioritization is key.

Unplanned Maintenance — maintenance task that occurs unexpectedly.

Scheduled Maintenance — Scheduled Maintenance is any task that is assigned to an xxx and scheduled. It can either be a recurring task done at regular intervals or a one-time task.

Periodic Maintenance — Periodic Maintenance refers to activities performed based on a set time interval.

Reliability Improvement — Reliability Improvement MTBF trending Up and Minor Stop trending down.

Maintainability Improvement — Maintainability Improvement Improvement in the ease of doing maintenance. Maintenance speed increase, frequency need decrease, visualization

A Planned Maintenance System has the following elements.

Planned Maintenance

Preventive Maintenance

The intent is to implement restorative maintenance to prevent failure. This is done using.

Scheduled Maintenance

Scheduled maintenance is any task that is given a deadline and assigned to a maintenance technician. It can be a recurring task done at regular intervals or a one-time task. The primary goals of scheduled maintenance are to reduce unplanned or breakdown maintenance, equipment failure, and maintenance backlog.

Periodic Maintenance

Periodic maintenance is a strategy that requires maintenance tasks to be performed at set time or use frequency intervals while the equipment is still operational. Similar to scheduled maintenance, periodic maintenance activities are planned ahead of time and are performed with or without signs of deterioration. The time or frequency is determined by experience over time. It is intended to help reduce the cost along the supply chain, and to prevent collateral damage to equipment that may occur with an unexpected failure

Planned Run to Failure

Planned run to failure is a maintenance strategy where maintenance is only performed when equipment has failed. Unlike unplanned & reactive maintenance, run-to-failure maintenance is a chosen strategy that is designed to minimize total maintenance costs.

Continuous Improvement Maintenance

1. **Plan:** Identify a maintenance improvement opportunity and plan the improvement.
2. **Do:** Implement the improvement.
3. **Check:** Use data to analyze the results of the improvement and statistically validate whether it made a difference.
4. **Act:** If the improvement was successful, reapply it.

Reliability Improvement

Increasing the line's reliability, increases your production throughput. Increasing production throughput increases the company profit. That increase in company profit is often even bigger than just the throughput gain, because reliable lines operate at lower costs.

Maintainability Improvement

Maintainability involves learning from experience in order to improve the ability to maintain systems or improve the reliability of systems based on maintenance practices.

The measure of maintainability is the **ease** with which maintenance activities can be performed.

This "ease" is achieved through; training, improved documentation, use of similar equipment across the production system, the practice or rapid changeover maintenance and increased planned maintenance.

Unplanned Maintenance

Unplanned maintenance is any maintenance task that occurs unexpectedly. This happens when there is no formal strategy or plan to address repair, replacement, or inspection before it's needed.

The use of Stability Check sheets by both the operating and maintenance teams will minimize the occurrence of unplanned maintenance.

Maintenance Compelling Business Need (CBN) Linkage

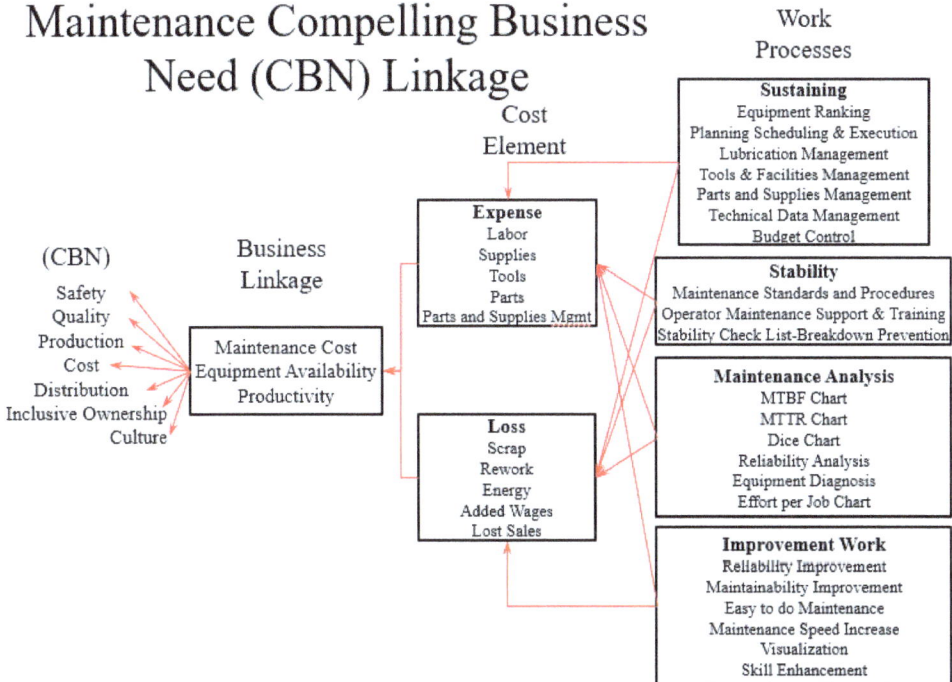

Work Processes

Cost Element

(CBN)
Safety
Quality
Production
Cost
Distribution
Inclusive Ownership
Culture

Business Linkage

Maintenance Cost
Equipment Availability
Productivity

Expense
Labor
Supplies
Tools
Parts
Parts and Supplies Mgmt

Loss
Scrap
Rework
Energy
Added Wages
Lost Sales

Sustaining
Equipment Ranking
Planning Scheduling & Execution
Lubrication Management
Tools & Facilities Management
Parts and Supplies Management
Technical Data Management
Budget Control

Stability
Maintenance Standards and Procedures
Operator Maintenance Support & Training
Stability Check List-Breakdown Prevention

Maintenance Analysis
MTBF Chart
MTTR Chart
Dice Chart
Reliability Analysis
Equipment Diagnosis
Effort per Job Chart

Improvement Work
Reliability Improvement
Maintainability Improvement
Easy to do Maintenance
Maintenance Speed Increase
Visualization
Skill Enhancement

The linkage to the business need is an element that the organization needs to emphasize to all personnel. The persons working on the equipment and ensuring the raw materials are converted to the desired product are the primary reason for a successful business.

The work performed must support the business.

Every line person has a maintenance role.

Maintenance Roles

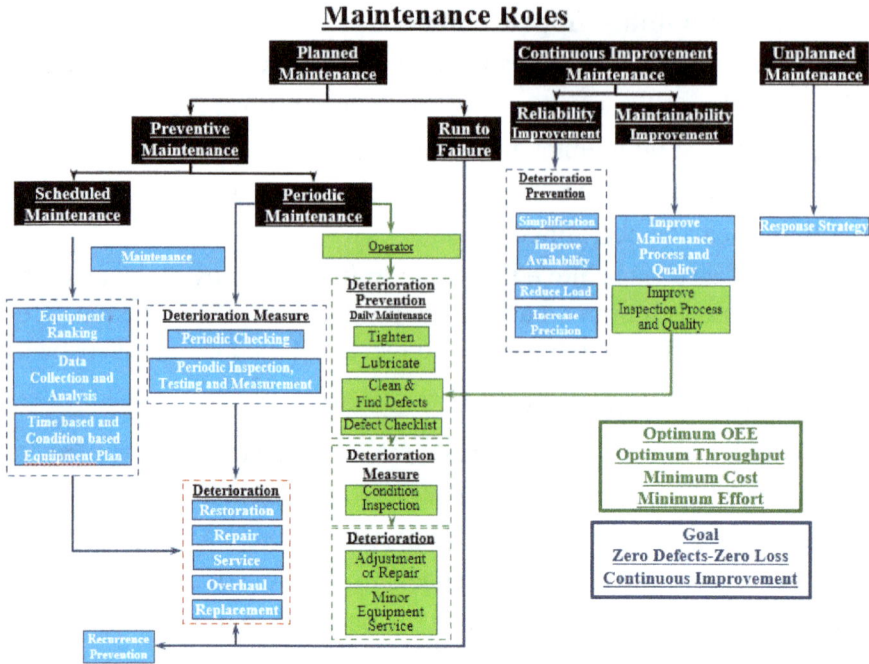

Everyone that interacts with the equipment needs to be involved in keeping it working as intended. The operator needs to understand the basic functions of the equipment.

Loose fasteners represent almost eighty percent of equipment minor stops. Minor stops represents a bigger loss than major breakdowns. Minor stops are often missed, or they require a slowdown of the equipment that often is not recorded as a loss. The first thing I have taught both operators and mechanics is the proper use of a torque wrench. This followed by the proper assembly of the bolt, lock washer and nut. These two learnings greatly reduce the minor stops.

Think of the equipment lubrication like it is blood in your veins. If it is clean, your heart will work for a long time. If the oil is clean the equipment will run for a long time.

The operator of the equipment should be responsible for maintaining basic lubrication. They should know how often to grease, what level to maintain oil levels.

The maintenance tech should be responsible for the more complicated grease and oil application. Complex Lubrication would be what might be associated with the size or the arrangement of the equipment. Lubrication requiring equipment disassembly would fall into this category.

Keep the line healthy and the work becomes easier.

Chapter 1: Maintenance Daily Management

Maintenance Daily Management

Maintenance daily management is by definition the management of the work done by maintenance team to meet the goal keeping the production equipment at its peak condition

Maintenance Daily Management Goal

The goal is to achieve zero product defects, zero material waste and zero lost equipment time.

It requires organizational coordination and adherence to standards.

Each operating team member has a secondary of doing their part to maintain the equipment.

The maintenance personnel are responsible to train the operators in the use of stability inspection check lists and in use of the clean, inspect and lubricate guides

Maintenance Daily Management

											"Drive Stability"								
	Maintenance Leader: Garrett Fortune												8/24/21		Maintenance Plan				
	Safety			Quality			Planned Maintenance		Unplanned	Stability					Cost	Maintenance Production Relationship			
	DOLA	Total Accidents	BOS	Foreign material	Product Weight	Defective Label	Scheduled Minutes	Achieved Minutes	Minutes	MTTR	MTBF	OEE/PR	Prior Month Machine Downtime	Current Month Total Machine Downtime	Maintenance Cost	Maintenance Method Based on Rank			
Maintenance	Thursday, October 1, 2020	2	1													Failure Logic and Terminology			
Product A	Maintenance												Hours	Hours	$	Plant Maintenance Priority Problems			
Line 1																			
Line 2																			
Line 3																			
Line 4																Throughput	Waste	Equipment	
Line 5																			
Product B																Plant Maintenance Priority Problems			
Line 1																Rank	100	80	60
Line 2																Location	Stuffer 1	Product B	Line 1
Line 3																Type	Environment	Waste	Equipment
Line 4																Brief Description			
Line 5																			
Product C																Improvement focus:			
Line 1																			
Line 2																			
Line 3																			
Line 4																			
Line 5																			

The daily work

Work orders

A maintenance work order provides details about maintenance, repair, or operations work, such as replacing a part, returning equipment to operating condition, or performing an inspection.

The purpose of a work order is to initiate a task, clarify what is to be done, specify completion dates, and give special instructions as needed.

There are different types of work orders. Some of these are.

- General Work Order.

  A general work order is any maintenance task that isn't considered a preventive maintenance, inspection, emergency, or corrective maintenance task.
- Preventive Maintenance work order
- Inspection work order
- Emergency work order
- Corrective Maintenance work order

Breakdown failure reports

A breakdown occurs when the equipment stops functioning. Breakdown is the result of failure and the effect that failure has. For example, if the temperature of your electric motor remains too high, it can melt the internals or it can cause the shaft to snap after the lubrication ceases to function, creating a breakdown.

Describe the failure in specific terms that clearly describes what you see not what may have happened. Get data that might help explain the effect over time and might show a trend. Review previous reports to see if the breakdown is repeating. Check for similar failures in other similar equipment.

Breakdown analysis

Mechanical attributes, spare parts, preventative maintenance procedures and employee skills are all analyzed and reworked to ensure the failure never returns.

Breakdown Analysis	Date:
	By:

Picture or Sketch

5 W-2H	
What ?	
When ?	
Where ?	
Which ?	
Who ?	
How ?	
How Much ?	

Summary

Why? Facts		
	Question	Answer
Why 1		
Why 2		
Why 3		
Why 4		
Why 5		

Summary

Action	Comment		Date
Immediate			
Longer Term			
Learning			Date
Maintenance Record			
Operator Standard			
Maintenance Standard			
OPL			

Common causes of equipment breakdowns, stops

- Frequent stops
- Improper operation
- Failure to perform or properly perform preventive maintenance
- Too much preventive maintenance
- Poor reliability culture

Breakdown Reasons

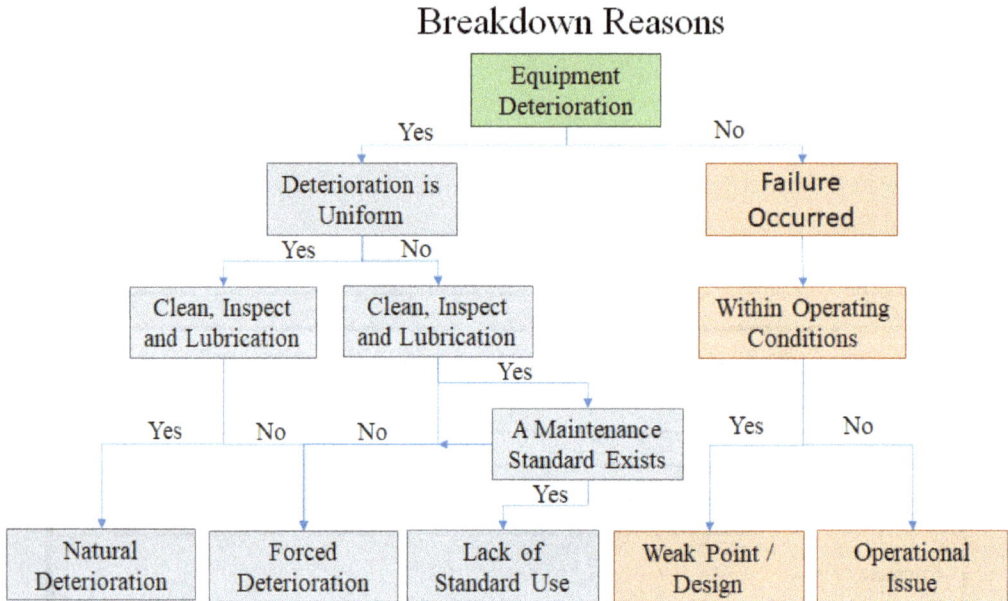

MTTR-MTBF reports

MTTR

The MTTR formula is calculated by dividing the total unplanned maintenance time by the total number of failures that the equipment experienced over a specific period.

The MTTR report summarizes the average downtime to repair.

MTTR is defined with the following formula:

MTTR = repair downtime / number of repairs completed.

MTBF

MTBF calculates the average period between two breakdowns. It is a measure of reliability. (how long the equipment typically works until it has a problem.)

It helps to make data-driven decisions on maintenance scheduling, safety, inventory management, and equipment design without relying on subjective observations.

Equipment records update

Equipment records are the primary data that is associated with the equipment. Each piece of equipment has a maintenance and performance record. The records establish the basic information such as:

- Identification number
- Description
- Account coding
- Dates of maintenance
- Location
- Status

It takes a great deal of time and effort to maintain each asset annually. Compounded by the total number of equipment and combined years of operation, documentation of maintenance tasks can easily get out of hand.

Having clear records of completed tasks can enable maintenance planners to plan out maintenance activities more easily and more efficiently delegate resources.

Keeping maintenance records updated is proof that assets are being taken care of.

Equipment maintenance logs provide data that can be analyzed. Different equipment can be compared by analyzing differences in maintenance costs incurred.

With available technologies, the benefits of having a well-documented equipment maintenance log are more easily achieved than ever.

Scheduling

Nearly all equipment needs some kind of regular maintenance and an equipment maintenance schedule. This type of maintenance is called planned maintenance (PM). Planned maintenance helps reduce the possibility of unexpected failure or repetitive equipment breakdowns.

An example maintenance plan is shown.

The goal is to maintain the line at its peak operating condition. For this to occur maintenance work needs to be closely coordinated with the operation. It is important to schedule major maintenance work, just as it is important to schedule product production.

The maintenance goal is to manage maintenance time in such a way as to optimize the production time.

The production line operator skills are improved to the point that they do the normal line lubrication and minor repairs. This is a key part of growing the production line skills. This operator hands on approach to line maintenance dramatically improves production line performance.

Maintenance Role and Partnership

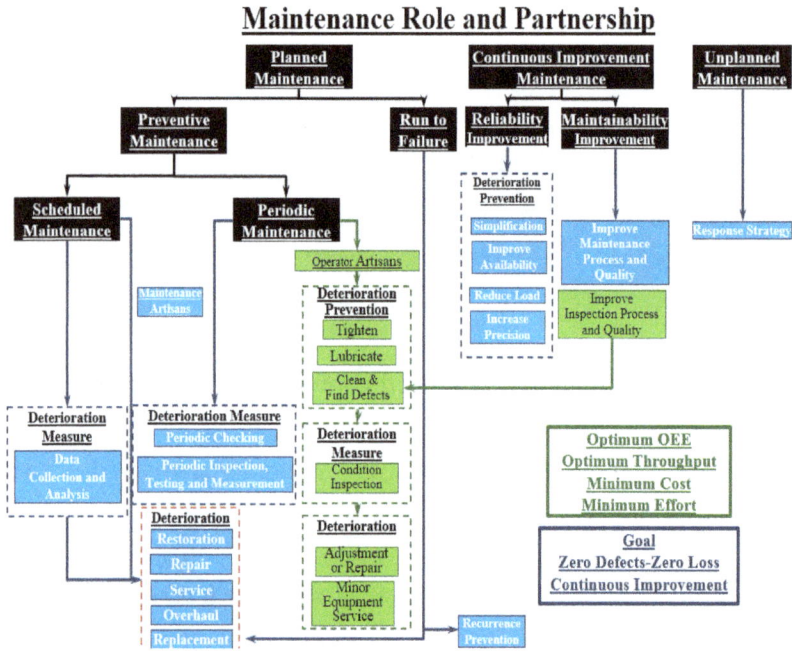

Backlog report

A backlog report highlights the work that needs to be completed. It is useful information to determine the effectiveness of the maintenance plan, and/or the capacity of the organization to implement the work

Maintenance Cost management

The cost associated with keeping equipment at its top performance by regularly checking it and repairing it is referred to as maintenance cost.

Minimizing cost is the goal of maintenance cost management. The best way to reduce maintenance costs is to prevent equipment malfunctions. Another key is to train the equipment operators and maintenance staff.

One of the most fundamental requirements of business operations is the ability to budget and control cost. Maintenance is a cost is often one of the larger costs of doing business for a production facility.

Spare parts management

The objective of spare parts management is to ensure the lowest overall cost of spare parts without compromising availability. A high level of availability requires spare parts to be available or for the delivery time to be as short as possible.

Key elements of spare parts management
1. Strategically identify all parts and to identify where they are located, and the retrieval time.
2. Utilize and Manage the Bill of Materials (BOM)
3. Simplify the work order process.
4. Centralize and consolidate parts.
5. Utilize a spare parts inventory control system.
6. Give every part a stock location so they can easily be located.

Parts and equipment Supply management

Parts management is the process that used to ensure that right spare part and resources are at the right place at the right time. It includes the purchase of physical goods, information, services, and any other necessary resources that ensure the production system is maintained at a high level of throughput.

Vendor performance report

A vendor performance report documents the good or bad performance of a vendor. Vendor performance can be managed by.
- Defining a vendor management strategy
- Defining performance criteria and expectations
- Collecting performance data and comparing them to the goals

Chapter 2: Stability Systems

A system is said to be stable if its output is under control. A stable system produces at a consistent output. Three key elements; Maintenance Standards, Zero Loss – Stability maintenance and the skill of the human are all critical in maintaining production system stability.

Maintenance standards and procedures (SMPs)

Maintenance of equipment is a repeating activity. This by definition means that a procedure can be written and optimized by those doing the work.

The persons writing the procedures should.

- have some training in writing SMPs
- someone involved with safety and environmental hazards should be involved.
- trained job maintenance personnel or subject matter experts should be involved.

The following information should be contained in a standard maintenance procedure.

- Formal title and document number.
- A statement to read the standard maintenance procedure before beginning work.
- Personal protective equipment (PPE) required to do the job.
- Safety and environmental hazards.
  Safety hazards are listed at the beginning of an SMP. Warnings should be repeated for each hazardous step.
  "Warning" to designate personnel harm and the word "Caution" to designate equipment harm.
- A task risk prediction should be the first step in use of the SMP.
  This is a risk prediction that links the risk to the counteraction resources.

Needed additional information for performing the procedure.
- A complete list of tools and materials.
- Listing of other documents needed.
- Needed photos and diagrams.
- Required measurements, standards, and tolerances.
- Required skill level.
- Required time.
- Required number of people.
- Required frequency.
- Required approval and review signatures.
- Space to provide feedback on the SMP's accuracy and effectiveness. Feedback is critical to the success of SMPs. In order for SMPs to be effective and accurate, a formal feedback mechanism should be supplied to the job performer. The SMP should be updated when feedback reveals mistakes or more effective ways to perform the job.

Writing Standard Maintenance Procedures

The SMP should focus on having just enough information to ensure the desired result.

There is no perfect SMP. The proper amount of detail will provide for trained persons to perform the job the first time.

The guiding elements for writing standard maintenance procedures are:
- The goal is to serve the user. The user must understand the procedure.
- Numbered line items and one item per step
- Keep wording short and precise.
- List steps in proper sequence.
- Use step check-offs.
- Whenever possible have the user enter quantitative values.
- Target sixth grade reading level.
- Use graphics to clarify meanings.
- Keep equipment and parts names consistent.
- Begin each step with a verb, action descriptor.
- For jobs with many steps, break the standard into sections.

Standard Maintenance Procedures are necessary to.
- protect personnel the health and safety.
- ensure that everyone performs a task to the same degree of precision.
- save time when performing a task.
- help ensure that standards and regulations are met.
- minimize the effects of personnel turnover.
- increase equipment reliability.
- serve as a training document.
- help document the equipment management procedure.
- help protect the environment.
- provide a basis for accident investigation.

Use Standard Maintenance Procedures

The goal is the consistent use of SMPs. Ensure they are part of the workorder process. Ensure they are periodically trained. Make SMPs 'easy to' documents. Post the SMP's on the equipment or at the operators station.

Zero Loss - Stability Check List application

Creating and utilizing specific detailed check sheets for the equipment provides a way to create stability and zero breakdowns. This approach has been successfully applied around the world and achieve significant results.

Operator maintenance and training

Training the personnel in the operation on the basic maintenance that they should perform is a key responsibility of the maintenance department. Skilled operators run stable operations.

—

Chapter 3: Sustaining and Support Systems

Sustaining

Sustaining systems are those systems that keep the production system healthy within the required performance boundaries. Each of these systems are described below

Equipment ranking

Equipment Ranking is a method used to assess the equipment's risk to maintaining the required production. The rank given to a piece of equipment is used to determine how often the equipment should be inspected or maintained and gives the maintenance scheduler a guide as to which work orders can be rescheduled to a future date, and which require more immediate attention.

Ranking Process

1. Develop the ranking criteria

   Use Production, Quality, Cost, Safety and Moral (PQCDSM) as well as maintainability and operability factors when ranking. Apply a risk prediction approach to properly rank these factors.
2. List all the equipment to be ranked
3. Have a team technically and business diverse team to the ranking.

Equipment	Rating Impact								Risk Prediction Number			Rank Value
	Production	Quality	Cost	Delivery	Safety	Moral	Maintainability	Operability	Occurrence Frequency	Ease of dectection	Impact Severity	
Equipment 1												
Equipment 2												
Equipment 3												
Equipment ...												

Note each participant gets to use 1-5 to rate each element. The group must agree on the rating for each element. The sum of the agreed to numbers is put into the column. The Rank Value is the multiplication of each value entered for each equipment rating category

4. List the equipment in rank order and determine the maintain action for each equipment.

Planning and Scheduling

Planning and Scheduling flows naturally from the ranking process. It is a critical part of maintaining the health of the production lines. It must be an integral part of the production planning process. Maintenance time should be a key part of the production plan.

Operation	Equipment		Maintence Plan 2020												Maintence Plan 2021											
		Jan	Feb	Mar	Apr	May	Jun	Jul	Aug	Sep	Oct	Nov	Dec	Jan	Feb	Mar	Apr	May	Jun	Jul	Aug	Sep	Oct	Nov	Dec
Operation 1																									
Department 1	Line 1	H	Major OH	s	s	s	s	s	s	s	s	s	s	H	s	s	Major OH	s	s	s	s	s	s	s	s
	Line 2	s	H	Major OH	s	s	s	s	s	s	s	s	s	s	H	s	Major OH	s	s	s	s	s	s	s	s
	Line 3	s	s	H	Major OH	s	s	s	s	s	s	s	s	s	s	H	Major OH	s	s	s	s	s	s	s	s
Department 2	Line 1	s	s	s	s	H	Major OH	s	s	s	s	s	s	s	s	s	s	H	s	s	Major OH	s	s	s	s
	Line 2	s	s	s	s	s	H	Major OH	s	s	s	s	s	s	s	s	s	s	H	s	Major OH	s	s	s	s
	Line 3	s	s	s	s	s	s	H	Major OH	s	s	s	s	s	s	s	s	s	s	H	Major OH	s	s	s	s
Department 3	Line 1	s	s	s	s	s	s	s	s	Major OH	s	s	s	s	s	s	s	s	s	s	s	s	H	s	Major OH
	Line 2	s	s	s	s	s	s	s	s	H	Major OH	s	s	s	s	s	s	s	s	s	s	s	s	H	Major OH
	Line 3	s	s	s	s	s	s	s	s	s	H	Major OH	s	s	s	s	s	s	s	s	s	s	s	H	Major OH
Operation 2																									
Department 1	Line 1	Major OH	s	s	s	s	s	s	s	s	s	s	s	Major OH	s	s	s	s	s	s	s	s	s	s	s
	Line 2	s	Major OH	H	s	s	s	s	s	s	s	s	s	s	Major OH	H	s	s	s	s	s	s	s	s	s
	Line 3	s	s	Major OH	H	s	s	s	s	s	s	s	s	s	s	Major OH	H	s	s	s	s	s	s	s	s
Department 2	Line 1	s	s	s	s	Major OH	H	s	s	s	s	s	s	s	s	s	s	Major OH	H	s	s	s	s	s	s
	Line 2	s	s	s	s	s	Major OH	H	s	s	s	s	s	s	s	s	s	s	Major OH	H	s	s	s	s	s
	Line 3	s	s	s	s	s	s	Major OH	H	s	s	s	s	s	s	s	s	s	s	Major OH	H	s	s	s	s
Department 3	Line 1	s	s	s	s	s	s	s	s	Major OH	H	s	s	s	s	s	s	s	s	s	s	Major OH	H	s	s
	Line 2	s	s	s	s	s	s	s	s	s	Major OH	H	s	s	s	s	s	s	s	s	s	s	Major OH	H	s
	Line 3	s	s	s	s	s	s	s	s	s	s	Major OH	H	s	s	s	s	s	s	s	s	s	s	Major OH	H

Major OH = Production Line Down for a minimum of one week
H = Production Line Down for 24 Hrs
s = Support as needed

Lubrication Management

Think of lubrication management as a blood transfusion process. The act of getting the oil from the lubrication system storage area to the equipment is that transfusion process. It must be clean, sufficient, and done at the proper frequency. Check the equipment manufacturer's guide for lubrication requirements.

It includes both the equipment operators and the maintenance personnel. It is the L in CIL (Clean, Inspect and Lubricate).

See Lubrication System to understand the design that allows the efficient management of lubrication.

Parts and supplies management

Think of this as having the required part quickly available. It may be in a location that is on site or economically close at hand. And the process of getting it to the line is predefined.

See Parts and Supplies System to understand the design that allows the efficient management of parts and supplies.

Tools and facilities management

Think of the surgeon holding out his hand as he calls out "scalpel". The right tool is immediately at hand, no waiting or fetching. The surgeon is in the operating room and the environment is controlled and the tools are readily available. Both operating and maintenance personnel should be able to function like a surgeon.

See Tools and Facilities System to understand the design that allows the efficient management of tools and facilities.

Technical data management

Technical data management is the process of obtaining, storing, organizing, and maintaining the data created and collected by the technical community. The technical data management process includes a combination of operational, maintenance and business functions that collectively aim to make sure that the data is accurate, available, and accessible.

This information impacts the maintenance planning process.

Shutdown maintenance

Shutdown Maintenance is maintenance that can only be performed while equipment is not in use. Shutting down machinery is costly, but sometimes due to the nature of the defective part/machine, shutdown maintenance is the only viable maintenance procedure.

This is a very common process for many production equipment.

Maintenance Budget Control

The Maintenance Budget is an operating budget that is set aside in a single fiscal year for maintenance activities on the organization's assets. The maintenance budget typically falls into the elements of the Planned Maintenance System.

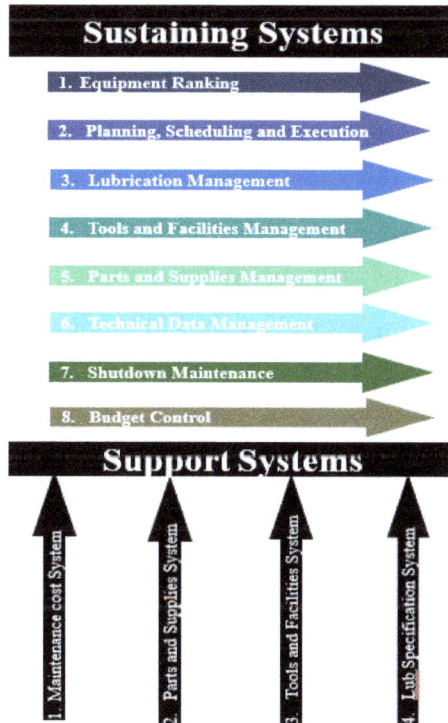

Support
Maintenance Cost System

A maintenance cost system tracks all the maintenance cost elements. This maintenance cost information is utilized in evaluating the equipment and the systems that support maintaining the equipment. It is a fundamental element in determining what the maintenance budget should be.

Parts and Supplies System

The Parts and Supplies System functions to ensure that both repairable and consumable replacement parts are available.

An example of a repairable part is the replacement of a dull cutting blade with a sharp one. The dull blade is then reconditioned to be available to replace the one that will become dull.

An example of a consumable part would be a battery, a light bulb or a motion sensor that gets replaced.

The part and supply system ensures that each type is available in sufficient quantities.

Tools and Facilities System

The Tools and Facilities System functions to ensure that the required tools are sufficient and are available at the point of use. It also ensures that the required facilities are available and are in superior condition.

Lubrication System

The lubrication system ensures that the required lubrication fluid is available.

It ensures that the lubrication is kept in the proper clean condition and is easy to use.

The system optimizes the number of lubricants and the method of dispersing it.

Chapter 4: Maintenance Analysis

Maintenance Analysis is a way to determine how to keep the production system in top shape. It is the method of determining how often the equipment will need attention in order to keep it in good working order.

There are a number of useful analysis tools.

Dice chart application

The dice chart is a visual bar chart of the mechanical, electrical, and pneumatic failures. It provides a visual way to share the problems a production line is experiencing.

Posting such a chart at the production line area communicates to the operations personnel the types of problems that are active on their line.

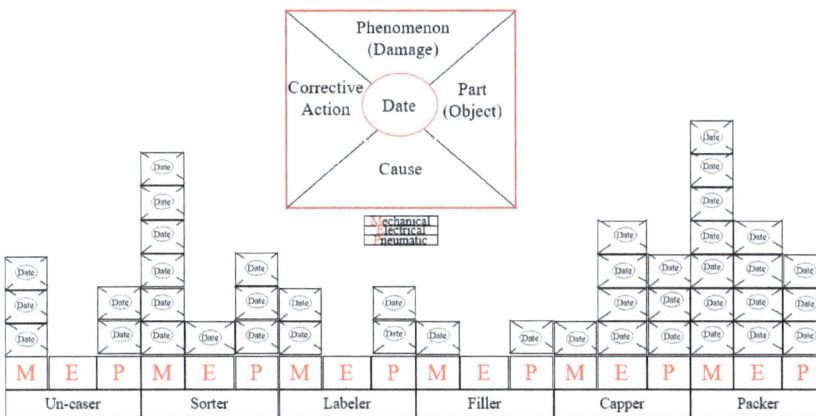

Dice Chart

Mean Time Between Failure (MTBF) Chart

The MTBF chart provides a visual way for all personnel to understand the performance of a line and the corrective actions that help to improve the MTBF.

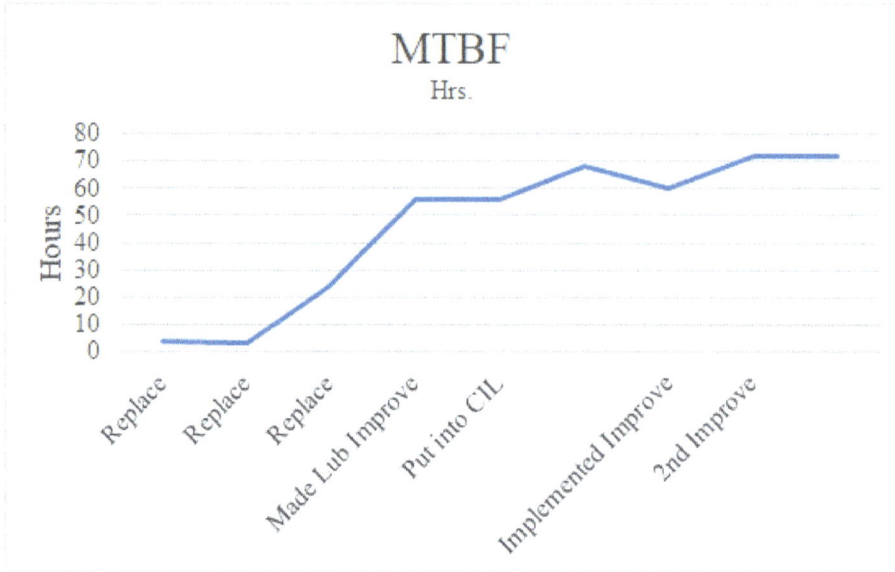

MTBF
Hrs.

Chart showing Hours (0-80) on the y-axis with x-axis categories: Replace, Replace, Replace, Made Lub Improve, Put into CIL, Implemented Improve, 2nd Improve. The line rises from near 0 up to about 70 hours.

MTTR Chart

The MTTR chart provides a visual way for all personnel to understand the performance of the maintenance actions that help to improve the MTTR. Improvement of MTTR often goes hand-in-hand with improved MTBF.

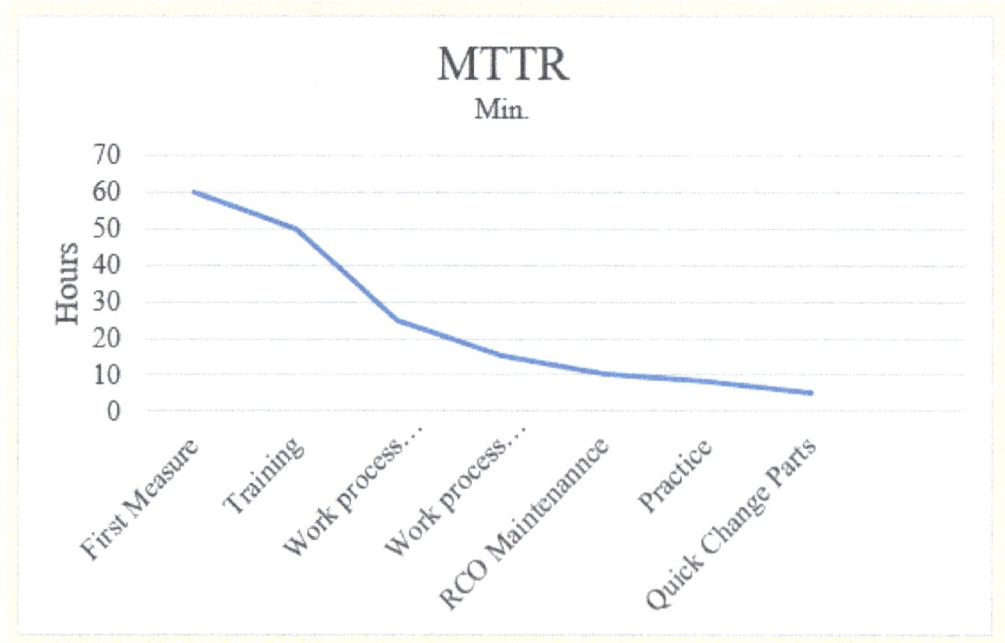

MTTR
Min.

Chart showing Hours (0-70) on the y-axis with x-axis categories: First Measure, Training, Work process…, Work process…, RCO Maintenannce, Practice, Quick Change Parts. The line declines from about 60 down to about 5 minutes.

Reliability Analysis

Reliability is defined as the probability that a production line or process will perform its intended operation for a specified length of time or will operate in a defined environment without failure.

Reliability of a line is calculated as follows.

$$PR = \frac{\text{Net Production}}{\text{Scheduled Time} * \text{Target Rate}}$$

Measurement Intent:

The intent of Process Reliability is to measure the extent to which Scheduled Time invested in Production, Maintenance and improvement activities is converted into Net Production. Some Scheduled time will be utilized to do maintenance and some scheduled time will be lost to unplanned downtime.

Target rate is the intended production target rate. By definition target rate is equal to or less then the ideal rate of the system.

The Reliability Constraint

The constraint is the point in the production system that limits the through put of the system. This point will move as the capability of the constraining point is improved. Ideally the system will be balanced, and a constraint will not be present. However, due to cost, most production systems will have a constraint designed in at the most expensive or costly production transformation.

The improvement focus for a production line should be the constraint and ensuring that other points in the process do not slow down, block or starve the constraint.

Analysis at the constraint

Schedule Utilization $_{Constraint}$ = Scheduled Time $_{Constraint}$ / Calendar Time

Rate Utilization $_{Constraint}$ = Target Rate $_{Constraint}$ / Ideal Rate $_{Constraint}$

PrReliability $_{Constraint}$ = Net Production $_{constraint}$ / Scheduled Time $_{Constraint}$ * Target Rate $_{Constraint}$

Capacity Utilization Constraint = $SU_{constraint}$ * $RU_{constraint}$ * $PR_{constraint}$

Equipment Diagnosis

The use of the Clean, Inspect and Lubricate CIL is a form of equipment diagnosis.

Taking equipment readings of oil level, temperature, vibration is the process of collecting data that allows for early detection of problems.

Determining what type of fault would explain the reading values of various sensors provides the basis for pre-emptive maintenance action.

The operations personnel play a major role in providing this early warning. It benefits them to have a line that is stable and runs continuously. It benefits the maintenance area since it allows for good maintenance planning.

Job Effort analysis

The purpose of job effort analysis is to provide data on how long a job takes. A comparison of how long this takes for an expert to do provides the basis for what others might need training on.

It also provides the basis for evaluating improvements to the work process.

The goal is to make daily work easier to do.

Cost Analysis

A cost analysis is the process of comparing the projected or estimated costs associated with a specific work process or processes.

The goal is to determine how the maintenance budget is being spent. The focus should first be on the big expenditures to see if there is a way to improve the spend situation.

For the maintenance on a production line, the focus should be on the equipment that generates the most cost. This information will aid in the determination of the action to take.

Minimizing over all maintenance cost is always a goal. It should be approached from the actionable level of making maintenance improvements.

Maintenance Measures

Equipment:
- Availability
- Maintainability

Maintenance Efficiency
- MTTR

Maintenance Effectiveness
- Downtime
- Operating Rate
- MTBF

Overall Effectiveness
- Availability
- Performance Rate
- Quality Rate

Chapter 5: Maintenance improvement and simplification

Maintenance improvement and simplification is a part of the maintenance daily management process.

Example of a Maintenance Daily Management board

									"Drive Stability"									
		Maintenance Leader: Garrett Fortune													8:24:21		Maintenance Plan	
		Safety			Quality			Planned Maintenance	Unplanned				Stability			Cost	Maintenance Production Relationship	
		DOLA	Total Accidents	BOS	Foreign material	Product Weight	Defective Label	Scheduled Minutes	Achieved Minutes	Minutes	MTTR	MTBF	OEI/PR	Prior Month Machine Downtime	Current Month Total Machine Downtime	Maintenance Cost	Maintenance Method Based on Rank	
Maintenance		Thursday, October 1, 2020	2	2													Failure Logic and Terminology	
Product A	Maintenance													Hours	Hours	$		

Plant Maintenance Priority Problems			
Rank	100	80	60
Location	Stuffer 1	Product B	Line 1
Type	Environment	Waste	Equipment
Brief Description			

Improvement focus:

The five Maintenance improvement issues are:
1. Lack of data quality in the PM notifications
2. Inability to capture required data in work orders
3. Incomplete information regarding Activity, Cause, and Damage.
4. Missing data in work orders for problem Analysis
5. Time consuming and high costs of maintenance training

Understanding the loss logic associated with various cause of line stops provides the basis for the action that needs to occur to minimize the loss.

Stops and Loss Logic

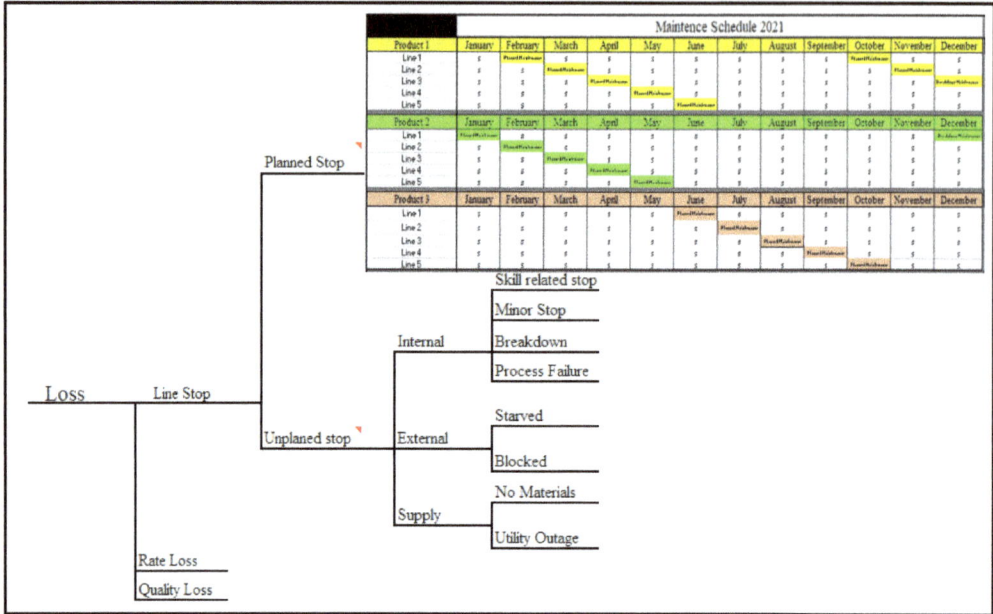

Maintenance improvement will require a close understanding of the work, the equipment, the interaction of the human with the equipment.

However, the focus on making small improvements everyday is a critical element that daily management empowers.

Ron Mueller P.E.
- Integrated Work Systems (IWS) materials author
- Coach to dozens of Manufacturing Directors across the world.
- Certified TPM Coach.
- Tested and proven to enable true breakthrough improvement of Supply Chains.

A proven leader of smart systems implementation across supply, manufacturing, 0and distribution to drive out cost, inefficiencies and to establish synchronized Supply Chains. He utilized the best thinking of Japan's TPM leaders and crafted the necessary related pillars and systems that work in Consumer Products Manufacturing. The results delivered include reduction of Raw and Finished Product Inventories by 40%. Delivered over $100 million is loss reduction through focused systems Workshops across dozens of sites. Developed P&G IWS program materials for external sale. Winner of P&G's Diamond Award for Contribution to Product Supply.

Core Competencies include:

- ✓ Coaching Manufacturing Leadership,
- ✓ Implementation of Integrated Work Systems,
- ✓ Statistical Replenishment design and implementation,
- ✓ Supply Chain Synchronization: author of 3 books in the Stress Free™ series that aid Business and Supply Chain leaders to develop and improve their organization's performance.

Gordon Miller P.E.

- Manufacturing Performance Program
- Development and Delivery Expert.
- Application of Intelligent Manufacturing technology against biggest business challenges with proven business results.

A record as a collaborative and leading-edge thinker, developing programs to deliver cost, productivity and growth enabling manufacturing technology systems deployed via smart standards and empowered teams. As an early developer of PR/OEE measures and improvement programs, has experience with unlocking organization capability for improvement with smart strategies. Led program that developed initial P&G Manufacturing Execution System, leveraged globally across multiple GBUs. Influenced Beauty and Household Care manufacturing systems changes that enabled and leveraged global standardization for rapid footprint growth. Experience that enabled 50% reduction in OEE losses. Experience as a leader of corporate STEM talent strategy can assess and devise approaches to ensure Talent needs for the challenging future are met.

Core Competencies include:

- ✓ Global Productivity Program Design and Management,
- ✓ Advanced Manufacturing Technology Innovation and Strategy Development,
- ✓ Development of Highly Effective Global Teams,
- ✓ Vendor development and management, Organization Capability Development,
- ✓ Talent Strategy

Design of the Stress Free Solutions Suite

Stress Free Manufacturing Solutions is available in two parts:

Stress Free Manufacturing Solutions – ***the*** *underline* ***book***

- Presents the theory
- It comes as an e-book or
- Seven by Ten Paperback copy or

Stress Free Manufacturing Solutions - ***Excel Workbook***

- Provides an "easy to document" environment
- This provides the user a way to optimize their resources and time.

Other Books by Ron Mueller

Stress Free™ Supply Chain Solutions

Stress Free™ Manufacturing Solution

Stress Free™ Work Process Solutions

Stress Free™ Changeover Solutions

Stress Free™ Daily Management Solutions

Stress Free™ Maintenance Solutions

Around the World Publishing LLC

4914 Cooper Road Suite 144

Cincinnati, Ohio 45242-9998

www.ingramcontent.com/pod-product-compliance
Lightning Source LLC
Chambersburg PA
CBHW081749200326
41597CB00024B/4445